# Table of
# Contents

# Welcome!

We're so excited to have you and your little one join us for this playful, hands-on science adventure!

Inside, you'll find simple and exciting activities that invite your child to observe, ask questions, get a little messy, and think like a real scientist—because guess what? They are scientists. And so are you. This workbook is designed to encourage fun, connection, and discovery. There's no pressure to do every activity—choose what feels right for your family and enjoy the journey together.

You don't need a lab coat or microscope to be a scientist—just curiosity, love, and a sense of wonder. Thank you for being your child's first and forever science partner.

**With joy and discovery,**
*Ms. B. & The Tiny Sparks Team*

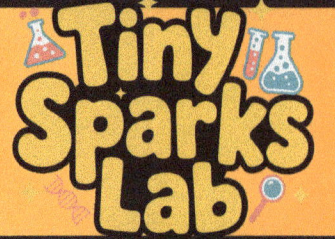

# Science Time Tips for
# Tiny Sparks

## Be Curious Together

Ask open-ended questions like:

> *"What do you notice?"*

> *"What do you think will happen next?"*

You don't have to know the answers—wondering together is what science is all about.

## Explore the Everyday

Science is everywhere— in your kitchen, bathtub, and even your backyard.

Point out patterns, changes, colors, and textures. Nature walks, water play, cooking, or even sorting laundry can become science time!

## Use Science Words

Encourage your child to use words like observe, predict, test, and conclude. It helps them feel—and think—like real professionals.

## Encourage Drawing & Talking

Invite your child to draw what they see, explain what they think, or act it out. Every scribble or silly explanation builds scientific thinking.

## Celebrate Mistakes & Surprises

If an experiment gets messy or doesn't go as planned, that's science too! Scientists make guesses, try things, and learn from every surprise.

## Capture the Moments

Take photos, record videos, or start a "Science Journal" with your child's work. These keepsakes show how much your tiny scientist is growing.

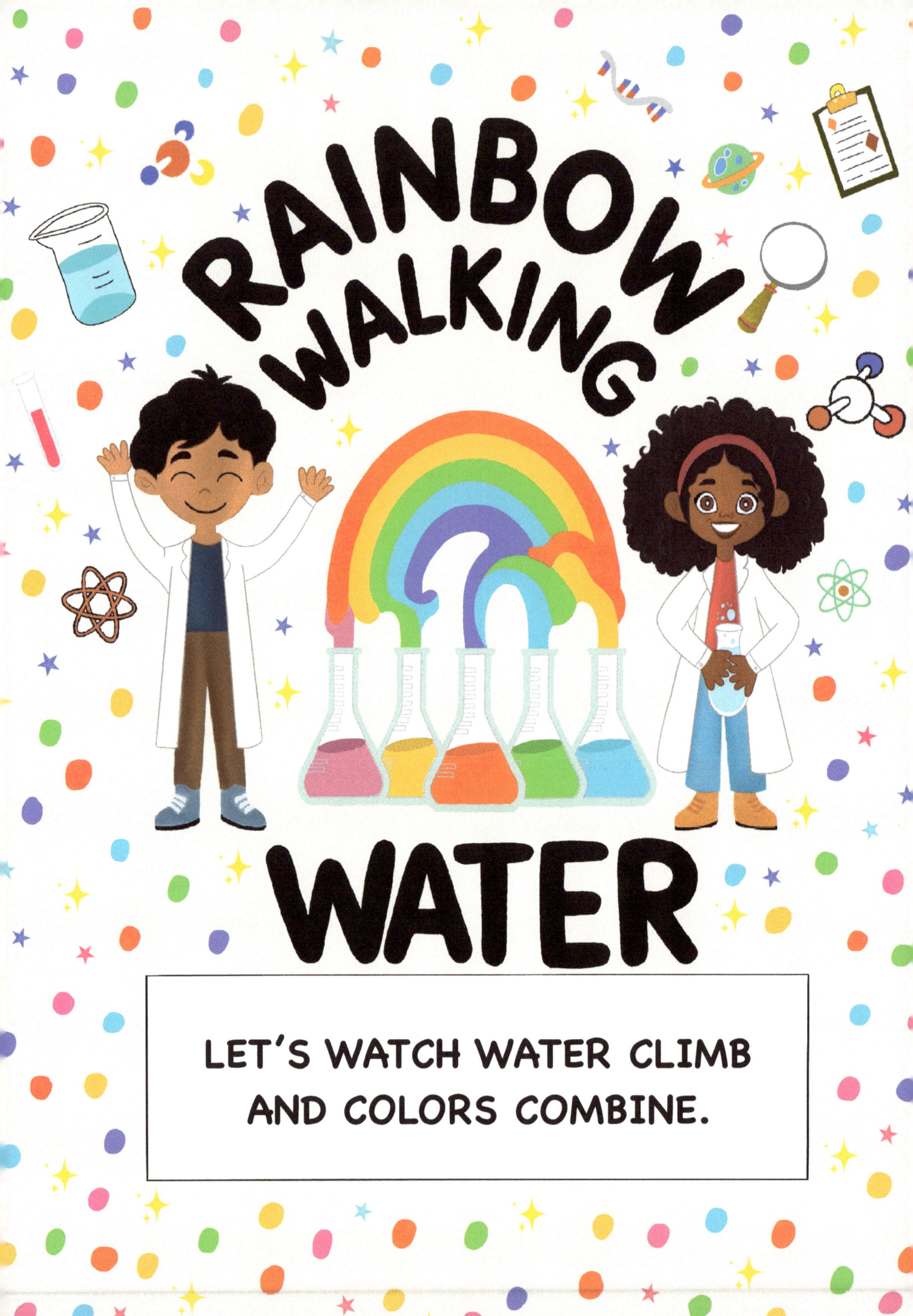

# RAINBOW WALKING WATER

LET'S WATCH WATER CLIMB AND COLORS COMBINE.

# Rainbow Walking Water

**SUPPLIES NEEDED:**

- 6 Clear plastic cups or jars
- Red, Yellow, and Blue food coloring
- Water

## Instructions

### 1. Arrange your cups

Line up 6 clear cups in a row. Fill cups 1, 3, and 5 halfway with water, leaving cups 2, 4, and 6 empty.

### 2. Add your colors

Put a few drops of red food coloring in cup 1, yellow in cup 3, and blue in cup 5. Stir each gently so the color spreads evenly.

### 3. Prepare the paper towels

Fold 3 paper towels lengthwise into strips. Each strip should be long enough to reach from one cup into the next.

### 4. Connect the cups

Place one end of a paper towel strip into the red cup and the other end into the empty cup next to it. Repeat between yellow and the empty cup, and between blue and the empty cup.

### 5. Watch the magic happen

Over the next 30–60 minutes, the colored water will travel ("walk") up the paper towels and into the empty cups, mixing to create new colors like orange, green, and purple.

Tip: For the best results, leave the setup undisturbed for a few hours and check back to see the full rainbow effect!

## Try This!

1. Test with shorter or longer paper towel strips—does it change how fast the colors move?
2. Use more than 3 colors and see what combinations you get.
3. Try using napkins, tissues, or coffee filters instead of paper towels.

# What is happening?

The water is climbing up the paper towels through something called capillary action—kind of like how plants drink water through their stems. As the colored water moves, it meets in the empty cups and mixes together, making new colors like orange, green, and purple!

# Let's talk about it...

- What new colors appeared?
- How did the water move without being poured?
- Can you think of other times you've seen colors mix?

- What happened to the paper towels as the water moved?
- What would happen if we used a different material instead of paper towels?

# Say it together!

*ACTIVITY AFFIRMATION*

*"We are scientists because we ask questions and stay curious!"*

# Rainbow Walking Water

## *Reflection*

Write your notes, ideas, and thoughts about the activity on this page.

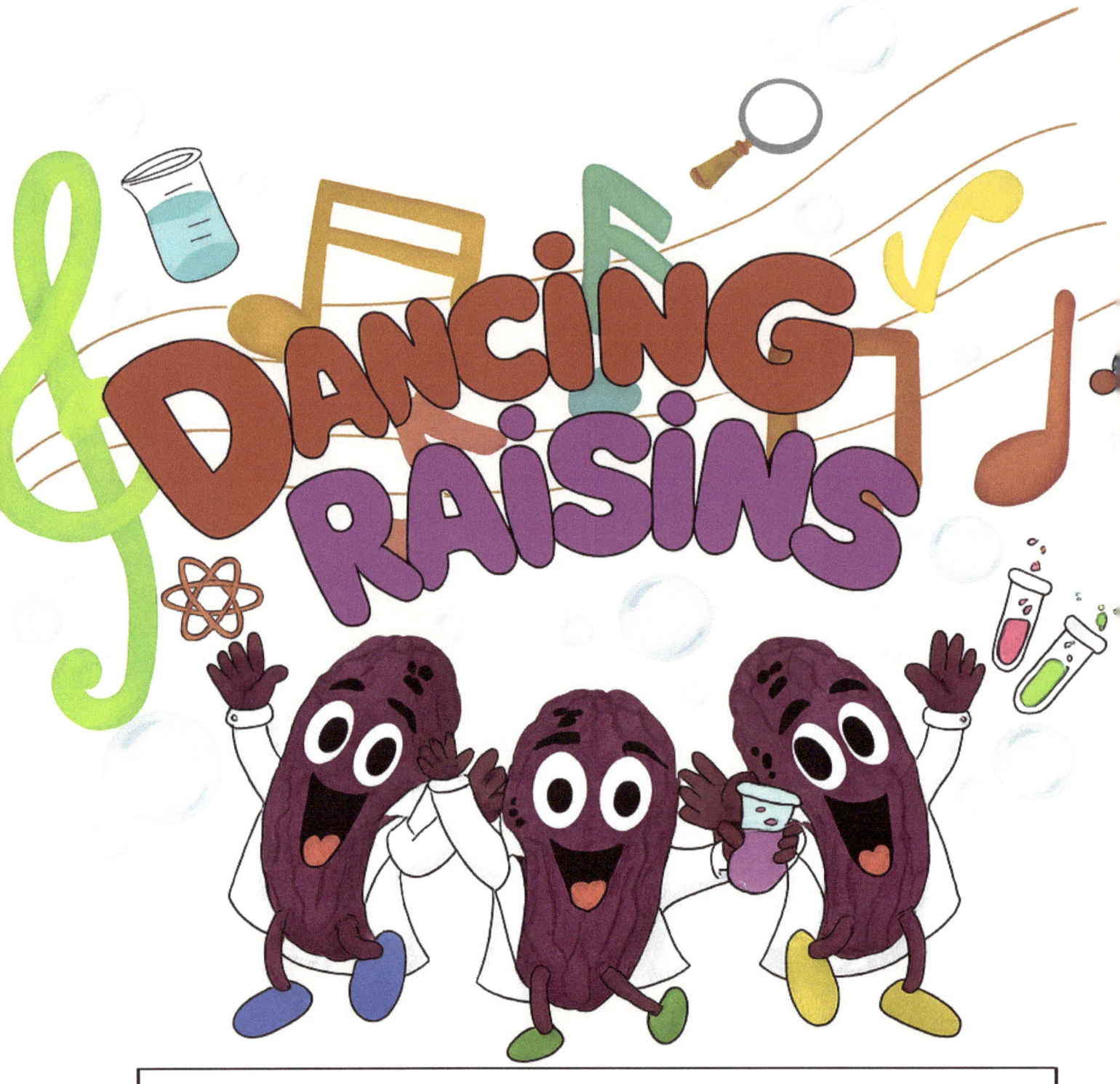

LET'S MAKE RAISINS
BOOGIE WITH BUBBLES!

# Dancing Raisins

## SUPPLIES NEEDED:

- 1 clear cup or glass
- A few raisins
- 1 teaspoon baking soda
- ½ cup vinegar
- Water
- A spoon
- A tray or towel (for easy cleanup!)

## Instructions

### 1. Fill your cup

Pour water into a clear cup until it's about halfway full.

### 2. Add baking soda

Stir in 1 teaspoon of baking soda until it dissolves completely.

### 3. Drop in the raisins

Add a few raisins and watch them sink to the bottom.

### 4. Make the magic happen

Slowly pour in ½ cup vinegar and watch for bubbles forming around the raisins.

### 5. Observe the dance

As bubbles lift the raisins to the top, they will pop and make the raisins sink again. Keep watching—they'll go up and down again and again!

## Try This!

1. Swap raisins for other small, lightweight foods—like corn kernels, blueberries, or bits of pasta.
2. Use a clear soda instead of baking soda and vinegar to see if it makes the raisins dance.
3. Try warm water instead of cold. Does it change the speed of the bubbles?
4. Add food coloring to make the dancing even more fun to watch.

DRY PASTA

BLUEBERRIES

CORN KERNELS

# What is happening?

When baking soda and vinegar mix, they make a gas called carbon dioxide. Tiny bubbles from the gas stick to the raisins and lift them up. When the bubbles pop, the raisins sink back down. Up and down they go— **just like they're dancing!**

## Let's talk about it...

- What did the raisins do at first, before we added vinegar?
- How did they change once the vinegar was added?
- Why do you think they went up and down?
- What did this experiment make you curious about?

- If you used something different (like corn kernels or blueberries), what happened?
- Did all objects "dance" the same way?
- Which objects floated the longest?
- What would happen if you used warm water instead of cold?

## Say it together!

— *ACTIVITY AFFIRMATION* —

*"We are scientists because we notice, wonder, and explore!"*

# Dancing Raisins

## *Reflection*

Write your notes, ideas, and thoughts about the activity on this page.

# Magic Milk Explosion

LET'S TURN PLAIN MILK
INTO RAINBOW ART.

# Magic Milk Explosion

## SUPPLIES NEEDED:

- Shallow dish or plate
- ½ cup milk (preferably whole milk)
- A few drops of liquid food coloring
- Small amount of dish soap
- Cotton swab or toothpick
- Paper towel or tray for cleanup

## Instructions

### 1. Pour the milk

Gently pour just enough milk into your dish to cover the bottom. You don't need much; just enough to make a thin, even layer.

### 2. Add the colors

Carefully drop in 3–4 spots of different food coloring near the center. Spread them out a little so the colors don't touch just yet.

### 3. Get your soap ready

Dip the tip of your cotton swab (or toothpick) into a small drop of dish soap. Make sure it's well-coated.

### 4. Start the magic

Touch the soapy tip right into the middle of the colors. Watch closely as the colors burst apart, swirl, and zoom around the dish.

### 5. Keep exploring

Try adding soap to a different part of the dish. Notice how the movement changes.

## Try This!

1. Test different kinds of milk (skim, 2%, whole, almond). Does the swirl change?
2. Add the soap in different spots—do the patterns look different?
3. Use a black background dish for a bolder effect.

# What is happening?

**M**ilk isn't just plain white liquid—it's full of tiny fat and protein particles floating in water. When you add dish soap, it zooms in and starts breaking apart the fat. This makes the milk's surface move and swirl. The food coloring doesn't mix with the fat, so it gets pushed around by the moving milk, creating bursts of color and swirling patterns. It's like a microscopic dance party happening right before your eyes!

# Let's talk about it...

- Which color moved first?
- How did the milk look before and after?
- What would happen if you used water instead of milk?
- How long did the colors keep moving?
- What name would you give your color explosion?

- What happened the moment the soap touched the milk?
- Did any colors mix together to make a new color?
- What happened when you added soap to a different spot in the dish?

# Say it together!

— *ACTIVITY AFFIRMATION* —

*"We are scientists because we try, even when it gets messy!"*

# Magic Milk Explosion

## Reflection

Write your notes, ideas, and thoughts about the activity on this page.

# SINK? OR FLOAT?

LET'S GUESS AND TEST WHAT FLOATS AND WHAT SINKS!

# Sink or Float?

## SUPPLIES NEEDED:

- Large bowl or clear container of water
- Variety of small household items (spoon, leaf, crayon, bottle top, toy, paperclip, rock, cotton ball)
- Towel for cleanup
- Paper and pen for data collection

## Instructions

### 1. Fill your container

Pour water into your bowl or clear tub until it's about three-quarters full. Leave a little room at the top to prevent big splashes.

### 2. Pick your test items

Gather a variety of small, safe objects from around the house or yard. Be sure to include items you think will definitely sink and others you think will float.

### 3. Make your predictions

Before putting each item in the water, guess what will happen. Will it sink to the bottom or stay on the surface?

### 4. Test each one

Gently drop each item into the water and watch closely. Does it bob at the top, hover in the middle, or sink right down?

### 5. Sort and compare

Group your floating items in one pile and your sinking items in another. Can you see patterns in what floats and what sinks?

## Try This!

1. Test the same objects in fresh water and salt water—what changes?
2. Test the same objects in warm water vs cold. Is there a difference?
3. Try shaping playdough into a ball vs. a boat to see which floats.

# What is happening?

Every object has something called density—how much "stuff" is packed into it. If an object is less dense than water (or if it has air trapped inside, like a boat), it floats. If it's more dense, it sinks. This is called buoyancy—the upward push water gives to things. Boats and beach balls float because they're filled with air or are less dense

than water. Rocks and metal sink because they're heavier for their size. By testing different objects, you're becoming a detective of what makes things sink or float.

# Let's talk about it...

- Which items surprised you?
- What do all the floating things have in common?
- How could you make a sinking item float?
- Which object floated the longest?

- Did any floating items eventually sink? Why do you think that happened?
- How did the size of an object affect whether it sank or floated?
- What would happen if you tested the same items in salt water?

## Say it together!

— ACTIVITY AFFIRMATION —

*"We are scientists because we test our ideas and see what happens!"*

# Sink or Float?

## Reflection

Write your notes, ideas, and thoughts about the activity on this page.

# BALLOON
# ROCKET RACE

ZOOM!
LET'S BLAST OFF WITH
BALLOON POWER!

# Balloon Rocket Race

## SUPPLIES NEEDED:

- 1 long piece of string (6–10 feet)
- 1 straw
- 1 balloon
- Tape
- 2 chairs or doorknobs
- Scissors

## Instructions

### 1. Thread the straw

Slide your straw onto the string so it moves freely back and forth.

### 2. Set up your track

Tie one end of the string to a chair, door handle, or sturdy object. Pull the string tight, then tie the other end to another object about 6–10 feet away.

### 3. Prepare your balloon

Blow up your balloon as big as you can without it popping, but don't tie it closed. Keep the neck pinched tight so no air escapes.

### 4. Attach the rocket

Use tape to secure the balloon to the straw, making sure the balloon opening points toward the end where you'll start.

### 5. Countdown and launch

Let go of the balloon neck and watch it race along the string! Try it again and see if you can make it go faster or farther.

## Try This!

1. Race two balloons at once.
2. Test balloon shapes—round vs. long.

# What is happening?

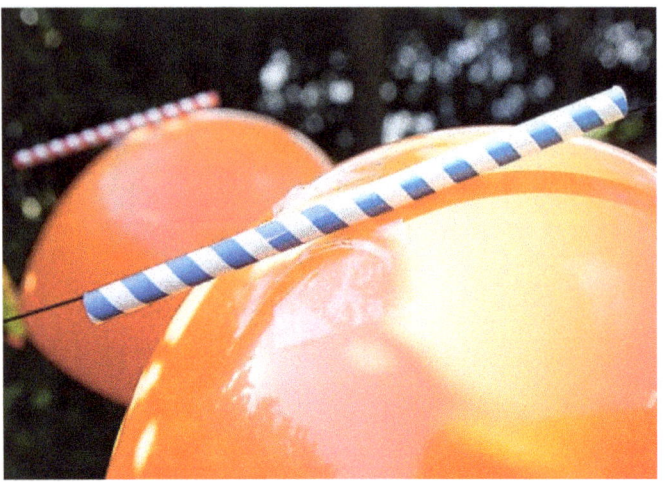

Balloons are like a little air tanks. When it's full, the air inside is under pressure, wanting to escape. When you let go, the air rushes out the back of the balloon. This push is called thrust—it's the same force that makes real rockets blast off into space! The straw and string act like a track to guide the balloon so it moves forward instead of spinning around. The more air you put in, the stronger the push, and the farther or faster it can go.

# Let's talk about it...

- How did the balloon move when you let go?
- How does balloon size affect speed or distance?
- What changes could make it go farther?
- Did the balloon go faster when it had more or less air?

- What could you change to make the rocket travel in a curve instead of straight?
- What other materials could we use instead of string for the rocket track?
- How would this work in space?

## Say it together!

— ACTIVITY AFFIRMATION —

*"We are scientists because we learn from mistakes and surprises!"*

# Balloon Rocket Race

## *Reflection*

Write your notes, ideas, and thoughts about the activity on this page.

# HOMEMADE
# PLAY DOUGH

LET'S BE KITCHEN
CHEMISTS AND DOUGH
DESIGNERS!

# Homemade Playdough

## SUPPLIES NEEDED:

- 1 cup flour
- ½ cup salt
- ½ cup water
- 1 tablespoon vegetable oil
- Food coloring (optional)
- Bowl and spoon

## Instructions

### 1. Mix your dry ingredients

In a bowl, combine 1 cup of flour with ½ cup of salt. Stir until they are fully blended.

### 2. Add your liquids

Pour in the water, vegetable oil, and a few drops of food coloring if you want colorful dough.

### 3. Stir and watch it form

Use a spoon to mix everything together until the dough starts to clump.

### 4. Knead with your hands

Sprinkle a little flour on your hands and squish, press, and roll the dough until it's smooth. If it's too sticky, add more flour; if it's too dry, add a drop or two of water.

### 5. Create and play

Shape your dough into animals, pretend food, or anything your imagination comes up with!

## Try This!

1. Make two batches with different amounts of salt—does one last longer?
2. Add glitter or scents for a sensory twist.

# What is happening?

When you mix flour, salt, and water, you're doing a simple kind of chemistry! The flour gives structure, the salt helps preserve the dough so it lasts longer, and the water makes everything stick together. The oil makes the dough softer and easier to shape. Once mixed, the ingredients change into something new—no longer dry powder or plain water, but a soft, squishy material you can mold into anything. You've just made a brand-new substance that didn't exist before!

# Let's talk about it...

- What does your dough feel like?
- How does adding more flour or water change it?
- How did the dough change as you mixed it?
- How does warm dough feel compared to cold dough?

- Did the texture change after you played with it for a while?
- What would happen if you left your dough out overnight?
- What shapes or designs can you make that stand up on their own?
- Can you make a science creature from your dough?

# Say it together!

—— *ACTIVITY AFFIRMATION* ——

*"We are scientists because we share what we see and think!"*

# Homemade Playdough

## *Reflection*

Write your notes, ideas, and thoughts about the activity on this page.

# MINI VOLCANO ERUPTION!

BOOM! BUBBLE! BLAST!
LET'S MAKE A VOLCANO
ERUPT!

# ACTIVITY
# Mini Volcano

## SUPPLIES NEEDED:

- Small plastic cup
- 2–3 tablespoons baking soda
- ¼ cup vinegar
- Squirt of dish soap
- Red/orange food coloring
- Tray or plate

## Instructions

### 1. Set the stage

Place your small cup or container in the middle of a tray or plate to catch the "lava."

### 2. Build your volcano (optional)

If you'd like, use play dough to shape a volcano around the cup, leaving the top open.

### 3. Add your ingredients

Pour 2–3 tablespoons of baking soda into the cup. Add a few drops of food coloring and a squirt of dish soap to make the lava more colorful and foamy.

### 4. Make it erupt

Slowly pour in the vinegar and step back! Watch as the fizzing "lava" bubbles out and flows down the sides.

## Try This!

1. Add more vinegar or more baking soda to see how the eruption changes.
2. Use warm vinegar for faster fizz.
3. Try lemon juice instead of vinegar.

# What is happening?

Baking soda is a base and vinegar is an acid. When they mix, they create a gas called carbon dioxide—the same gas we breathe out and that makes soda fizzy. The gas forms bubbles, and when you add dish soap, the bubbles stay longer, making foamy "lava." The food coloring makes it look even more like a volcano. Real volcanoes erupt when hot, melted rock (magma) and gases escape from deep inside the Earth. Your mini volcano uses a safe chemical reaction to mimic that explosive effect!

# Let's talk about it...

- How could you make a bigger eruption?
- What other colors could lava be?
- How does this compare to a real volcano?
- How many seconds did the eruption last?

- Did the bubbles move fast or slow?
- What happened when you added more vinegar after the first eruption?
- What sounds did your volcano make?
- How far did your "lava" flow?

# Say it together!

*ACTIVITY AFFIRMATION*

*"We are scientists because we look closely and pay attention!"*

# Mini Volcano

## *Reflection*

Write your notes, ideas, and thoughts about the activity on this page.

# SHADOW DRAWING

LET'S CHASE THE SUN AND
TRACE THE SHADOWS!

# ACTIVITY

# Shadow Drawing

**SUPPLIES NEEDED:**

- Sunny spot or lamp
- Toys or objects
- Paper
- Pencil, crayon, or marker

## Instructions

### 1. Choose your light

Find a sunny spot by a window or go outside, or set up a bright lamp indoors.

### 2. Place your object

Put a toy, block, or other item where the light hits it so a shadow falls clearly on your paper.

### 3. Trace the shadow

Hold your pencil, crayon, or marker steady as you follow the outline of the shadow.

### 4. Move and compare

Shift your object or light source and see how the shadow changes. Trace it again to compare shapes.

### 5. Decorate your shadow art

Turn it into a monster, superhero, or something silly!

# Try This!

1. Try to trace your objects in the same spot, but at different times of day.
2. Try to trace your or your science partner's shadow.
3. Layer objects to create combined shadows.

# What is happening?

Light travels in straight lines until it hits something solid. When that happens, the light can't go through, and a dark shape—called a shadow—is made behind the object. The size, shape, and direction of the shadow depend on where the light is coming from. If the light moves, the shadow changes too. Outside, the sun's position changes throughout the day, making shadows longer, shorter, or even disappear for a while. You're exploring how light, objects, and movement work together.

# Let's talk about it...

- How did the shadow change when you moved the object?
- What happens when the light is closer or farther away?
- How did the shadow's shape change when the light moved higher or lower?
- Did some objects make sharper shadows than others? Why?

- What happened when you used more than one light source?
- How does your shadow look in the morning compared to the afternoon?
- Could you make a moving story using just shadows?

## Say it together!

——— *ACTIVITY AFFIRMATION* ———

*"We are scientists because we create, imagine, and discover new things!"*

# Shadow Drawing

## Reflection

Write your notes, ideas, and thoughts about the activity on this page.

HELP! OUR TOYS ARE
TRAPPED IN ICE—LET'S SAVE
THEM USING SCIENCE!

# Ice Rescue Mission

## SUPPLIES NEEDED:

- Small toys
- Ice cube tray or containers
- Water
- Freezer
- Spoons, droppers, warm water, salt

## Instructions

### 1. Trap the toys

Place small toys into the compartments of an ice cube tray or muffin tin. Fill each space with water.

### 2. Freeze overnight

Carefully place your tray in the freezer and let it freeze until solid.

### 3. Begin the rescue

Pop the frozen toy-filled ice cubes out onto a tray or towel.

### 4. Test your tools

Use spoons, droppers of warm water, or a sprinkle of salt to try and melt the ice.

### 5. Compare methods

Notice which tool works fastest and which takes the longest. Keep going until all your toys are free!

## Try This!

1. Use colored water to freeze the toys.
2. Try warming the ice with a magnifying glass to see if it's faster.
3. Time each method to see which is quickest.

# What is happening?

When water gets cold enough—at or below 32°F (0°C)—it freezes into ice. The molecules in the water slow down and lock together into a solid. When ice warms up, the molecules speed up again and turn back into liquid water—this is called melting. Salt makes ice melt faster by lowering its freezing point, which is why roads are salted in winter. In this experiment, you're using tools and temperature changes to melt the ice and rescue the trapped toy. You're exploring states of matter and how to change them!

# Let's talk about it...

- How did the toy look through the ice before you freed it?
- Which melting method worked fastest?
- How long did it take to rescue your toys?
- Which melting tool was the easiest to use?

- How did salt change the ice?
- Did the ice melt faster in your hand or in warm water?
- What would happen if you froze the toys in layers of ice with different colors?

## Say it together!

— ACTIVITY AFFIRMATION —

*"We are scientists because we keep trying until we understand!"*

# Ice Rescue Mission

## Reflection

Write your notes, ideas, and thoughts about the activity on this page.

# Nature Color Hunt

**SUPPLIES NEEDED:**

- Color chart or paper with color spots
- Bag, basket, or hands

## Instructions

### 1. Get your chart ready

Use your color chart or make your own by drawing circles in different colors.

### 2. Pick your hunting ground

Go to your yard, a park, or even look out the window.

### 3. Search for matches

Look carefully for something in nature that matches each color on your chart—like a yellow flower, a red leaf, or a brown stick.

### 4. Mark your finds

Point to the color, circle it, or draw the item next to the color on your chart.

### 5. Share your rainbow

Show a family member your colorful finds and talk about which ones were easiest or hardest to discover.

## Try This!

1. Compare colors in different seasons.
2. Hunt for shades of just one color.

**Use the sheets on pages 41-42 to search for colors!**

# Indoor color Hunt

Can you find something of each color?
Write, draw or place your findings inside the colored circles!

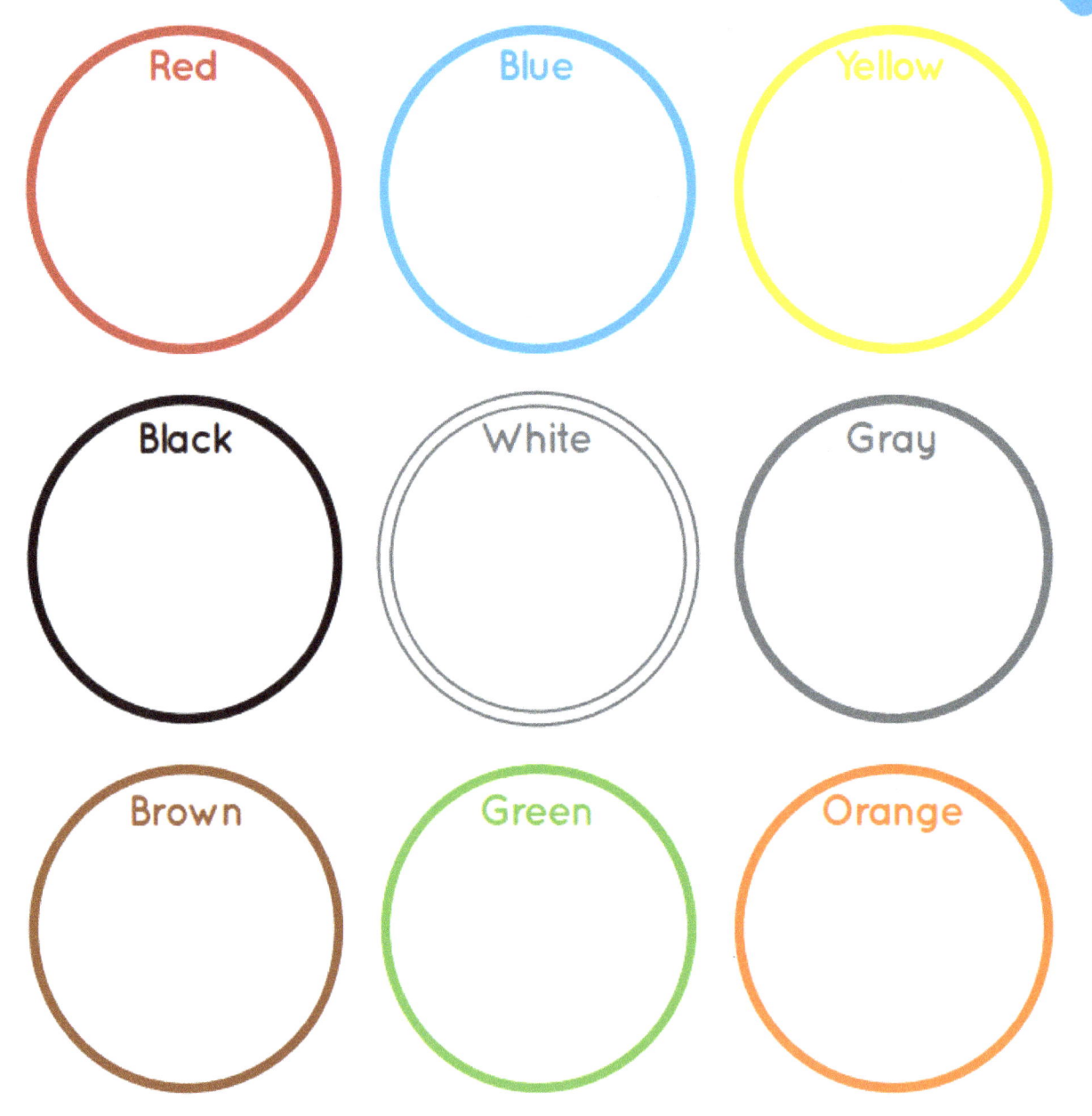

Red

Blue

Yellow

Black

White

Gray

Brown

Green

Orange

# outdoor color Hunt

## Can you find something of each color?
Write, draw or place your findings inside the colored circles!

Red

Blue

Yellow

Black

White

Gray

Brown

Green

Orange

# What is happening?

Nature is full of color, and each color can tell us something about the plants, animals, and seasons around us. Bright green leaves mean a plant is full of life, while brown or yellow leaves might mean it's fall or the plant is drying out. Flowers use bright colors to attract pollinators like bees and butterflies. Even the sky changes color depending on the time of day and weather. By looking closely for colors, you're practicing observation skills just like real scientists do when studying the environment.

# Let's talk about it...

- Which colors did you find the most of?
- Which color was hardest to find?
- Did any items match more than one color?
- What colors did you find that aren't on your chart?

- Did the same color look different in the sunlight vs. shade?
- Which colors reminded you of a season (spring, summer, fall, winter)?
- Did anything surprise you?

# Say it together!

—— ACTIVITY AFFIRMATION ——

*"We are scientists because we explore the world all around us!"*

# Nature Color Hunt

## *Reflection*

Write your notes, ideas, and thoughts about the activity on this page.

 # CONGRATULATIONS!

## You've Earned Your
## Official Scientist Badge!

**OFFICIAL SCIENTIST**

This badge is awarded to **YOU** for being a curious, creative and courageous scientist!

You should be PROUD of yourself!

**We are PROUD of YOU**

# Keep the Science Adventure Going!

**Science doesn't stop when the activity ends**—it's all around you, every day! Here are some fun ways to keep exploring at home:

- Ask Wonder Questions — Why does toast pop up? Why do bubbles float away? Every "why" is the start of an experiment!

- Be an Observer — Notice the shadows on the wall, the bugs in the yard, the patterns in the weather, or the magic in your kitchen.

- Experiment Anywhere — Mix, pour, and play with things you already have—water, sand, food coloring, or even everyday objects.

**Science time can be anytime—you just need your curiosity!**

## Parent Tips: What's Next?

1. Stop by your local library and discover children's books all about science adventures.

2. Head outdoors with a notebook and start your very own nature journal.

3. Try simple kitchen science—make butter, mix up slime, or watch rainbow colors climb through celery!

## AND ALWAYS, HAVE FUN EXPLORING!

**Did you capture the fun?**
*If you took photos or videos of your experiments, we'd love to see your tiny scientists in action! Share your moments on social media—just follow us and tag your post with **#TinySparksLab** so we can celebrate your discoveries together.*

**Follow us on Facebook & Instagram:**
**@TinySparksLab**